五个石头的故事

孔祥庚 著

云南出版集团
云南人民出版社

图书在版编目（CIP）数据

五个石头的故事 / 孔祥庚著. -- 昆明 : 云南人民出版社, 2016.4
ISBN 978-7-222-14123-0

Ⅰ. ①五… Ⅱ. ①孔… Ⅲ. ①化石 - 介绍 - 澄江县
Ⅳ. ①Q911.727.44

中国版本图书馆CIP数据核字(2016)第082228号

出 品 人：刘大伟
项目策划：赵石定
责任编辑：赵石定
王　韬
装帧设计：舒　宏
责任校对：郑燕燕
责任印制：洪中丽

五个石头的故事

孔祥庚　著

出版	云南出版集团　云南人民出版社
发行	云南人民出版社
社址	昆明市环城西路609号
邮编	650034
网址	www.ynpph.com.cn
E-mail	ynrms@sina.com
开本	889mm×1194mm　1/32
印张	3.625
字数	52千
版次	2016年4月第1版第1次印刷
印刷	云南出版印刷（集团）有限责任公司 云南新华印刷一厂
书号	ISBN 978-7-222-14123-0
定价	18.00元

如有图书质量及相关问题请与我社联系
审校部电话：0871-64164626　印制科电话：0871-64191534

云南人民出版社公众微信号

自　序

无论是自然科学，还是社会科学，无论是哲学、文学，还是宗教，其艰辛探索研究的最终目标便是——生命本源。

我们开启人类的智慧宝典《易经》，细细研读六十四卦，三百八十四爻，四千零九十六卦之，无一不是在追寻宇宙的本源，无一不是在追寻人类生命的本质，无一不是以万物之象说人议事，最后的结果却是：未济！宇宙未济！人生未济！宇宙从哪里来？人从哪里来？先有男人还是先有女人？一切皆未济矣！现代科学家在帽天山发现的无数古生物化石，证实了《易经》的伟大，回答了《易经》未知的一些基本问题。聊将五件标志性的古生物化石铭之为传，遴选一篇老故事为其佐证，命名为《五个石头的故事》，与读者共享东方神秘主义解密的乐趣。是为自序。

Preface

There is an ultimate goal of arduous explorations for research in both the natural and the social sciences, especially in the form of philosophy, of literature, of religion, which is the origin of life.

Zhouyi (*the Book of Changes*), as an extremely precious book on human wisdom, elucidates the origin of universe and the essence of human life through everything in the nature. By a careful reading of sixty-four hexagrams, three hundred and eighty-four yaos and four thousand ninety-six changes, it turns out that the final hexagram is the Weiji hexagram (the symbol of unfulfillment). Everything is an unfulfillment in the origin of universe or life. What came first, the chicken or the egg? Modern scientists discovered countless fossils in Maotianshan Mountain, which prove the extraordinary *Book of Changes* and answer the basic questions to it. I wrote about five symbolic fossils and an old story into a biography, named it *Five the story of the stone*, and this is my preface to share the fun of interpreting Eastern traditions of mysticism.

目　录

Contents

纳罗虫化石

帽天山何在？地图坐标：东经 102°58'38"，北纬 24°40'08" 。这里的抚仙湖深不可测，像流动的翡翠一样呵护着起伏峥嵘的群山。一座像圆顶草帽的山峰矗立于抚仙湖东北岸，冒顶上白云缭绕，如通苍天。不知何年何月，人们称它为帽天山。如此山水，印证了宝典的一语：“天一生水，地六成之。”宇宙万物生成的第一个元素是水。近几十年来，科学家们从帽天山页岩中发掘出数十万件古生物化石标本，其第一件古生物化石即纳罗虫

化石。

话说1984年，中国科学院南京地质古生物研究所的青年研究人员侯先光，千里迢迢而来，登上帽天山，剥土取石，剖面探秘。他渴饮山泉水，饥食冷馒头，与当地村民苦乐数十个昼夜。7月1日，流火突变小雨，闷热加泥泞，湿漉漉的石块剖面十分醒目可爱。下午3点许，侯先光挥动着小锤，突然从潮湿的页岩中连续敲出3块寒武纪时期的节肢动物化石。第一块化石是长尾纳罗虫化

帽天山

石：半圆形的头甲，椭圆形的尾甲，腿肢对称分布，似乎在游泳，像活的一样。头部有3对双分支附肢，尾甲上有22对双分肢附肢。他双手捧着长尾纳罗虫化石，兴奋不已，彻夜无眠。

却说，侯先光何以如此？他在南京攻读硕士研究生时就从有关资料中了解到：1909年，法国地质学家德普瑞特与满苏首次在中国澄江发现寒武纪岩层和化石。可惜，他们遭人诬陷为“编造化石”而失去了在印度支那地质调查所的职位。1942年，何春荪深入帽天山勘察磷矿时，找到了微型节肢动物化石，并将帽天山页岩层命名为“帽天山页岩”。抗日战争时期，中山大学地质系迁到云南澄江帽天山附近，从耶鲁大学归国的中山大学

1984年7月侯先光在帽天山考察途中

地质系教授、古生物学家杨遵仪，把这类微型节肢动物化石鉴定为高肌虫。侯先光教授穷源竟委，寻觅勘探，圆了无数科学家的百年梦。

且说纳罗虫名字的由来。它是美国古生物学家维尔卡特命名的。1909 年 8 月底的一天，维尔卡特带着考察队员和妻子、儿子在加拿大布尔吉斯峡谷找化石，考察结束返回时，妻子骑的马被一块石头绊了一下。他立即下马劈开那块石头，一枚保存软体附肢的动物化石奇迹般地出现了：这就是寒武纪中期的一枚节肢动物化石。后来，维尔卡特把在布尔吉斯峡谷采集的其中一枚节肢动物化石命名为纳罗虫。加拿大布尔吉斯页岩动物群很快被科学家称为“生物天堂”，维尔卡

1984 年 7 月侯先光在澄江采集化石

特也被称为19世纪伟大的传奇科学家。遗憾的是，布尔吉斯发现的60多种节肢动物化石，由于被压成纸片状而没有一件能够揭示其完整的腿肢。就说纳罗虫化石吧，帽天山的长尾纳罗虫、刺状纳罗虫化石都是肢体完好的标本，布尔吉斯的纳罗虫化石虽然与帽天山的相似，但是晚了1000多万年，不可能揭示地球上最古老的动物特征。纳罗虫化石在帽天山古生物化石群中，虽然属于比较多见的动物化石，但是科学界公认它是地球上最早的动物化石！

俗话说：“外行看热闹，内行看门道。”化石这东西，一般人拿在手里只是个毫无用

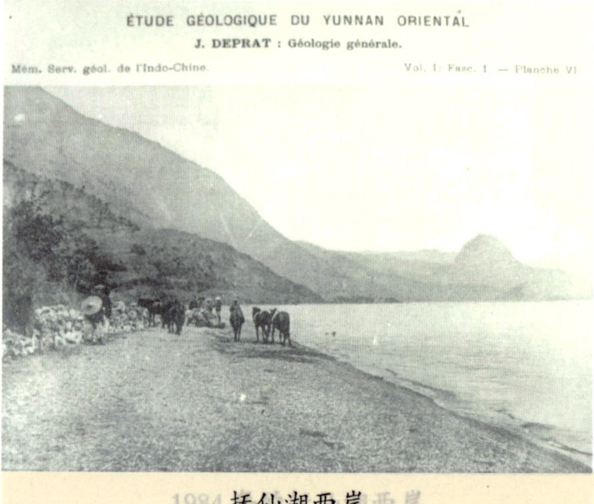

抚仙湖西岸

处的破石头，而在行家眼里却是历史长河中生命进化的直接证据。地球上的化石与人的成长相似，由于环境不同、时代不同，得到的机遇也不同。动物化石得以保存的机会极少，几乎只是动物数量的万分之几，动物的软体部分或没有硬骨骼的动物更难以保存为化石。帽天山首次发现的纳罗虫化石属于保存完好的软躯体节肢动物化石，随后又发现了许多举世罕见且保存完好的软躯体化石。可以说，帽天山纳罗虫的问世，引出了许多有意义的推测：地球是太阳系中唯一有复杂生命存在的行星，距今大约有 46 亿年。5.3 亿年前是寒武纪早期，地球上生活的各个动物门类几乎都在此时期从不同的地点同时大爆发。大约 4 亿年前出现鱼类，3 亿年前

抚仙湖旧景

出现爬行动物，2 亿年前出现哺乳动物，200 万年前出现人类。

帽天山第一枚纳罗虫化石的发现者侯先光也名声大震。有的领导人与知名人士曾经建议侯教授：“尽快转向产业化，与市场经济接轨，利用化石搞一点赚大钱的买卖。”游说者不少，侯教授却一声不吭。他私下对我说：“咱们只看得懂石头，看不懂金钱。”这话耐人寻味：古今中外沉迷于石头里炼丹、淘宝、赚大钱的人多如牛毛、数不胜数，但是真正能读懂石头这部无字天书的人，简直是凤毛麟角。女娲炼石补天，但她是神仙而不是凡人。春秋时期楚国的卞和，将在山中得到的一块璞玉献给厉王，被判为以石充玉欺骗君主而断了他的左脚；后来送给继位

2004 年落实温总理“三个保护”

的武王，又被以欺君之罪断了他的右脚；最后送给文王，才被认为是价值连城的和氏璧。卞和可谓中国历史上能读懂石头的英雄之一，但和氏璧只是供帝王将相赏心悦目的玩物，远远不及西方启蒙运动以来从石头里发现各种化学元素符号的科学家们，更比不上那些能从石头里看懂未来世界的科学家。侯先光在帽天山古生物化石群里发现的科学故事，比伊甸园神话更精彩美丽，这些化石证据进一步丰富了达尔文的生命演化论。以此推测，如果美国的维尔卡特是世界上第一个真正能读懂石头的伟大科学家，那么中国的侯先光应该是世界上第一个从石头里揭开地球生命奥秘的伟大科学家。

2006 年 10 月李纪恒到帽天山调研

有人说：“‘伟大’二字太夸张，他俩能够凭一片小石头震惊世界，完全是一种偶然的巧合。”这话没有错，但也没有完全对。维尔卡特偶然发现布尔吉斯古生物化石时，已经年近 60 了。之前，他在北美采集了 40

2006年10月李纪恒到帽天山调研

多年的寒武纪化石。侯先光偶然发现帽天山古生物化石之前，也曾踏破铁鞋四处考察。偶然成功之中隐藏着兴趣、勤奋、知识，三者俱全，才可能碰上运气，否则，遇到最精美的化石也会从自己的眼皮底下溜走。崇尚科学，首先应该崇尚科学家追求真善美的英雄气概。著名作家、诗人郁达夫讲得好："没有英雄的民族是可怜的民族，有了英雄而不知爱惜是无可救药的奴隶之邦。"中华民族的伟大复兴在呼唤科学，无数科学家在众里寻他千百度中崛起。侯先光教授已经步入灯火阑珊处：他先后荣获国家自然科学一等奖、何梁何利基金科学与技术进步奖、全国"五一"劳动奖章、全国先进工作者、云南省自然科学特等奖、云南省杰出贡献奖等若干奖项和荣誉。我聊以纳罗虫化石抒怀，口占一绝：

大地洪荒数亿年，
小虫演化隐绵延。
弯弯曲曲重重雾，
译破天机敬大贤。

2006 年拜访侯先光教授

Naraoia

Where is Maotianshan Mountain? Map coordinates are located at 102°58'38" east longitude and 24°40'08" north latitude. Here the Fuxian Lake is blessedly deep surrounded by the towering mountains, of which the slowly flowing water looks as green as jadeite. A mountain stands in the northeast of the Fuxian Lake, looking like a round straw hat. Flowing mists weave themselves around the peak, as if it were standing out against the sky. It's hard to know since when people have called it Maotianshan Mountain. Such natural poetry recalls the book saying: Water is the source of all things of creation. The first element of cosmic inventory is water. In recent decades, scientists have found the very first fossil — *Naraoia* from hundreds of thousands of fossils from Maotianshan Shale.

In 1984, Hou Xianguang had crossed thousands of miles to climb the Maotianshan Mountain to collect materials. By that time he was a young researcher in Nanjing Institute of Geology and Palaeontology, Chinese Academy of Sciences. He took cold steamed bread for food and spring water for drink, kept working several ten days with local villagers. On July 1st, the weather was rainy and hot. The locality was muddy but the stones were very conspicuous because of wet. At around 3 o'clock on that muggy afternoon, Hou split the slab and beheld a creature unlike any he had seen. He found three specimens of Cambrian arthropod fossils in the wet shale. The first one is *Naraoia longicaudata*: oval or semi-circular headshield, a tailshield longer than the cephalon, symmetric limbs, three pairs of biramous legs in the head, and twenty-two pairs of biramous legs in the body. It looked so alive that seemed like swimming. Hou Xianguang was too excited to fall asleep all night, holding the specimens of *Naraoia longicaudata* with both hands.

Why was Hou Xianguang so heartfelt excited? He learnt from the relevant materials when he was a graduate student in Nanjing: In 1909, French geologists J.Deprat and H.Mansuy found Cambrian stratum and fossils for the first time in China. Unfortunately, they were falsely accused of 'made up fossils' and lost their positions at Indo-China geological survey. In 1942, He Chunsun found micro-fossils of arthropods when he was doing phosphorous mineral exploration in Maotianshan Mountain, and named the shale bed Maotianshan Shale. At the anti-Japanese war period, Department of geology, Zhongshan University moved to Maotianshan area of Chengjiang County, Yunnan Province. Yang Zunyi, professor in the Zhongshan University's department of geology, who was graduated from Yale, identified these micro arthropods as bradoriida. Professor Hou Xianguang made unremitting efforts and his exploration realized a great many scientists' dream of a hundred years.

Naraoia was named by American paleontologist Walcott. In late July of 1909,

Walcott took his teammates, his wife and son to find fossils at Burgess Valley in Canada. While they were on their way back, Walcott's wife fell off her horse because the horse tripped over a stone. He dismounted and split the stone, and found a miraculous fossil with soft tissues, which was a kind of arthropod in Middle Cambrian. Walcott named this species *Naraoia*. The slabs of Burgess Shale that Walcott excavated were what scientists call "the heaven of livings". Some hailed Walcott as "the Greatest Scientist in the 19th Century". It was a pity that Burgess Shale contained over 60 species of arthropods but they didn't reveal a complete appendage/limb because of the preservation. For example, *Naraoia longicaudata* and *Naraoia spinosa* are both well-preserved in Maotianshan Shale. *Naraoia* in Burgess Shale is similar in shape but over 10 million years later than that of Maotianshan Shale. Characteristics are best revealed in Maotianshan Shale of ancient animals. *Naraoia* is common species in Maotianshan Shale, but the world scientific community recognized it as the earliest animal fossils on the earth.

The fossils on the earth and the human have a similar way in growth. Because of the dissimilar times and environments, they have dissimilar chances. There are extremely few opportunities for the preservation of animal fossils, especially for the soft part or animals without skeletons. After arthropod *Naraoia*, there are many different kinds of animals being found in Maotianshan Mountain, which are very rare and exquisite with soft bodies. In a matter of speaking, *Naraoia*, which came out of Maotianshan Shale, drew forth to lots of meaningful speculations: Of all the planets in our own solar system, we are now pretty certain the Earth is the only one on which life can survive, with a 4.6 billion-year history. The Maotianshan Shale, which is about 530 million years old of Early Cambrian, came soon after a period of time known as the Cambrian explosion, when most major groups of complex animals arose over a surprisingly short period. Fish are first found about 400 million years ago, while reptiles appear roughly 300 million years ago. Mammals also appear in the fossil record about 200 million years ago. Humans became human, as it were, with the emergence 2 million years ago of a species called *Homo erectus*.

As the saying goes, dilettante watch the scene of bustle, adept the contents. The fossil is generally considered as a useless stone, but in the experts' eyes, it is the direct evidence of evolution of life in the Earth's history.

Hou Xianguang, who discovered the first *Naraoia* specimen in Maotianshan Shale, has gained fame. Some celebrity once suggested Prof. Hou, "Transform into industrialization, connect with market economy, and make use of fossils to make a fortune". Despite many lobbyists, Prof. Hou remains silent. He said to me privately "We only know how to observe stones", which is very thought provoking. In history, there are countless people addicted to making profits from stones, but only few people can truly understand the intangible book. Goddess Nu Wa can mend the sky with stones, but she is a goddess not a human. Bian He in Chunqiu Era consecrated an uncut jade to the

Emperor Li, who sentenced him of fraud and cut his left leg; he consecrated to Emperor Wu, who sentenced him of fraud and cut his right leg; finally he consecrated to Emperor Wen, who thought it was a priceless jade. Bian He is one of the heroes who know stones in Chinese history, but the jade is only for emperors to entertain, far less from the stones in which scientists since the western enlightenment have found various chemical elements, and even far less from those in which scientists can see the future. Prof. Hou's story is more amazing than the Garden of Eden, because these fossils further enrich Darwin's evolution. Therefore, we can estimate that if Walcott from the United States is the first scientist to truly understand stones, Hou Xianguang from China is the first to unveil the myth of life from stones.

Some may think "Great" is too exaggerating, because it is completely coincidence that they can shock the world by a tiny stone. It is not accurate. Walcott has been 60 when he discovered Burgess Fossils. Before that, he had been collecting the Cambrian fossils for more than 40 years. Before accidentally discovering Fossil Specimens in Maotianshan Mountain, Hou Xianguang had been looking around persistently. The accidental success results from interest, diligence and knowledge. All the three elements make luck possible. Otherwise, one can't discover the most delicate fossils even in front of his nose. To respect science is first to respect scientists' heroic spirit of pursuing the true, the good and the beautiful. Yu Dafu, a famous poet, said "A nation without heroes is miserable; a nation without respect toward heroes is an incurable slave". The Renaissance dream of Chinese has been summoning science, and numerous scientists have been rising. Prof. Hou has been close to the summit: he gained several awards, such as the first prize in Chief of National Natural Science, Science and Technology Progress Prize by Ho Leung Ho Lee Foundation, National Labor Medal, The National Advanced Worker, Natural Science Prize in Yunnan, Yunnan Outstanding Contribution Award and so on. I made a quatrain of *Naraoia* to express my feeling：

It's been billions of years since the primitive earth began,

That the worms are building on a design of evolution.

The evolutionary roads have twists and turns,

Waiting for the people of great talent and virtue to explore.

三叶虫化石

话说三叶虫化石：背部特征十分明显，两条纵沟将两肋分开，像三片树叶匀称拼合而成，故命名为三叶虫化石。它是世界上分布最广、化石种类发现最多的海洋无脊椎动物化石。目前有记载的三叶虫化石已达17000多种，由于澄江动物化石群中的三叶虫诞生于寒武纪初期，因此仅有4种三叶虫，其中最著名的就是寒武纪早期三叶虫中的始莱德利基虫。我们生活的滇池、抚仙湖畔寒武纪地层中经常可见三叶虫化石，农户家里、小学生书包里也偶然能看到三叶虫化石，玩石市场上更是屡见不鲜，花上几百块钱即可买到一枚三叶虫化石。它虽然是一枚极普通的化石，但是它的经历却极不平凡。侯先光教授曾经讲过它精美的故事。

澄江动物群

寒武纪时期，我们今天生活的滇中大地还是古生代海洋，几乎没有陆地，没有飞禽，寂然荒凉，天水之间只有原始海洋里的生物。三叶虫诞生于寒武纪早期，属于原始海洋节肢动物，有的科学家把它称为节肢动物的祖先。古人云："祖者，始也，己所从始也。自父之父以上皆得称焉。"祖先之名，冠之不易。三叶虫诞生之际，身长不过数厘米，是原始海洋中披硬壳的节肢动物。由于它善于在恶劣的原始海洋中求生存、敢于在弱肉强食的竞争中谋发展，因此，它腿部发达，眼睛较大，最大者体长可达1米多，外壳格外坚硬，有的外壳上还长出锐刺，故成为原始海洋里非常丰富多彩的强大家族。

天地万物的奥秘一字概之：变！三叶虫经过2.8亿年的拼搏，其进化已经走到了极点，终于在地球环境的急剧变化中，随着古生代的结束而灭绝。三叶虫诞生得威武，灭绝得悲壮！它开创了寒武纪时代，结束了漫长的古生代，从寒武纪初期到二叠纪末期，经历了整个古生代，见证了整个古生代的沧桑巨变。这个漫长的渐变，是优胜劣汰的过程，是适者生存的过程，是具有生存能力的生命代替丧失生存能力的生命的过程。人类应该借鉴古生物的演变过程，小小心心地走好每一步，认认真真地过好每一天。聊以小诗自励：

本是无生变有生，
人生终要向归程。
百年路上须珍重，
步步三思稳健行。

首发点展厅

书归正传。1909 年，法国地质学家 Deprat 和 Mansuy 出版的《滇东地质状况备忘录》就记载了在澄江采集的，百年来一直作为寒武纪早期最著名的三叶虫化石："云南头虫"和始莱德利基虫。"云南头虫"和始莱德利基虫是澄江动物化石群中的重要成员之一，始莱德利基虫还被古生物学家作为划分和对比地层的标准化石。1958 年 4 月，中国邮电部发行三枚中国古生物特种邮票，

首发点内景

其中就有三叶虫化石。这枚三叶虫化石邮票选至山东上寒武统中部长山组的篙里山虫化石，标志着中国在研究地球早期生命和地球历史的成就和地位，也寄寓着我们的情怀。

三叶虫虽然是低级动物，它的演化过程却达到了最高层次：竞争。达尔文提出生物优胜劣汰规律之前，我们的老祖宗孔子早在《系传》中阐述过："方以类聚，物以群分，吉凶生矣。"吉，就是优胜；凶，就是淘汰，任何生物种类与群体都有凶吉。因为，时代不仅仅属于某一个生命，而是属于所有的生命。交臂非故也！机遇属于自己时，不应该轻易放过；机遇不属于自己时，不能勉强。天命难违！三叶虫逃脱不了天命。人生如是：

三叶虫化石邮票

命运像一双无形的手，既捉弄你，又恩赐你。无形的右手，在生老病死上恩赐平等；无形的左手，在名利情欲上捉弄无常。芸芸众生，在这双无形的手下，或消极沉沦，或怨天骂地，或养精蓄锐，或拼搏奋起。人类历史上曾经出现过许多仁人志士，他们累垮不倒下，饿死不低头，劫难不后退，屡败屡战，不胜不休。“盖文王拘而演《周易》；仲尼厄而作《春秋》；屈原放逐，乃赋《离骚》；左丘失明，厥有《国语》；孙子膑脚，兵法修列；不韦迁蜀，世传《吕览》；韩非囚秦，《说难》《孤愤》；《诗》三百篇，大抵圣贤发愤之所为作也。”这些古代圣贤在无形的命运之手下，遵循了人类社会的规律。司马迁弘扬了他们的高尚品格，成为与厄运抗争的伟人，是人类应对命运的成功典范。三叶虫遵循了宇宙规律，是低级动物应对命运的佼佼者之一。

古海折腾三叶虫，
英姿但见能拼搏。
人生何处无风险，
苦去甜来更欢乐。

Trilobite

Speaking of trilobite, it has very clear dorsal features. Trilobites are so named because the segments on their dorsal surface usually possess longitudinal furrows that form a three lobed division of the body, just like three pieces of leaves combining together. It is the oceanic invertebrate who is most widespread in the world and whose traces are found in the widest range of fossils, until now, there are more than 17,000 species of trilobite fossils recorded. Since the trilobite fossils in Chengjiang Fossil Site was born in the early stage of Cambrian, there were only four species of trilobites found, the most popular of which is the *Eoredlichia* that can date back to the early Cambrian period. The trilobite fossil was often found in the Dian Lake and in Cambrian along the Fuxian Lake where we live; sometimes we can see it even in farmer's house and in the schoolchild's school bag; The trilobite fossil in the stone market is nothing new now, it only costs you a few hundreds to take one home. It is an ordinary fossil but with no ordinary life experience. Professor Hou Xianguang once explained their extraordinary experience.

In Cambrian period, the place (Middle Yunnan) we live today was a Paleozoic marine, nearly no land, no fowls or beasts, lonely and desolate, there was only organisms living in the primitive ocean. Trilobites originated in early Cambrian period, belong to primitive ocean arthropods, some scientists considered it as the ancestors of the arthropod. Ancient saying goes, "Ancestors are what we originate from, those who born our grandfather can be called so." The credit such as "ancestors of the arthropod" is not something we just give, when trilobite was firstly born, it was only centimeters long and it was the arthropod who lives in Primitive Ocean and has shell on the back, but in order to survive in the Primitive Ocean of severe environment and countless predators and competitors, they developed muscular legs, big eyes, longer body — the longest one is up to 1 meter, sharp thorns on the exceptionally hard shell, and in the end they became the powerful and versatile family in Primitive Ocean.

The profound mystery of the Universe rests on one word: change! With a 280-million-year of endeavor, the evolution of trilobite came to an end; at last it became extinct with the termination of the Paleozoic. Trilobites born with credit and extinct with heroic. They establish the Cambrian period, put an end to the long Paleozoic, born in the early Cambrian, go as far as Late Permian, survived the whole Paleozoic and witness the great changes in that period. This is a lengthy and gradual change, a process of survival of the fittest, a process that the creature loss of viability will be replaced by those with viability. We should learn from the evolution process of ancient biological and take every step seriously, seize each day. The following poem can be seen as self-motivation:

Life comes from the germ in the Ocean,
But it is always heading to its destination.
We must treasure the time we have,
Make each decision wise, each step steady.

Go back to the trilobite. In 1909, French geologist Deprat and Mansuy's book *Memorandum of eastern Yunnan Geology* recorded the trilobite fossil collected in Chengjiang — *Yunnanocephalus* and *Eoredlichia*, which was the most famous fossils of early Cambrian in the past hundreds of years. *Yunnanocephalus* and *Eoredlichia* are also the significant member in the Chengjiang Fossil Site, and in paleontologist's mind, the *Eoredlichia* is the standard fossil that can distinguish and compare the stratum. In April of 1958, the three special stamps of China's paleontology issued by Chinese Ministry of Posts and Telecommunications is trilobite fossil. The trilobite fossil on these three stamps is the *Kaolishania* in the Changshan Formation (central upper Cambrian), Shandong, which indicated the achievement and status of China's research in early life and history of the earth.

Trilobites are inferior animals, but their evolution reaches the highest level: competition! Before the theory of natural selection was propounded by Charles Darwin, our ancestor Confucius stated in his *System* (*Xi Zhuan*), "Creatures of the same kind come together, those auspicious will survive, ominous disappear". Auspicious here means those with advantage will win, ominous here means those without will lose. Winning and Losing exist in all biotic communities, because the era belongs to all living creatures, not just one. There is no such thing as permanent, we should seize the opportunity when we have the chance, and not cry on it when it is not belonging to us. Everything has its destiny, trilobite has no exception, and so is life: Life is like the invisible hands, they teases you as well as blesses you. The invisible right hand favors equality in birth, senility, illness and death; the invisible left hand favors mysterious in fame, wealth, affection, and lust. All mortal living beings are under the effect of these invisible hands, some despair, some grumble, some consolidate and some strive. There were many people with lofty ideals in human history who never fall out when they are completely worn out, never surrender when they are starving, never flinch when they are suffering a catastrophe, instead they keep on fighting until they get success. *Zhouyi* was wrote when King Gai Wen was detained; *The Spring and Autumn Annals* was wrote when Confucius was in his arduous years; *Li Sao* was composed when Qu Yuan was exiled; *Guo Yu* was completed even Zuo Qiu was blink; Sun Zi's kneecaps were removed, but he still finish the Warcraft; Lv Buwei was banished to Shu, but he completed *Lv Shi Chun Qiu* (Lv's *Commentaries of History*); *Shuo Nan*, *Hu Fen* and *Shi* was done by Han Fei when he was imprisoned in Qin, all these works are achieved by sages who were in their difficult times. These ancient sages follow the human being's regulations. Si Maqian, the historian in Western Han period, carried forward their distingue character and became the greatest man who fought against adversity, and set a successful example for human being against miserable destiny. Trilobite follows the universal rule and become one of the best in dealing with destiny among inferior animals.

Trilobites were tested to the utmost in Primitive Ocean,

They fight, survived and developed.

Life alike is full of risks,

But bitter brings sweet to more joy.

抚仙湖虫化石

话说侯先光教授发现帽天山动物化石群之后，将两枚亲手采集的动物化石标本命名为“抚仙湖虫”，从而引起好奇者猜疑：究竟先有抚仙湖虫化石，还是先有抚仙湖？这犹如“先有蛋还是先有鸡”一样怪哉！也许是为了解开这道怪题，中央电视台两次对抚仙湖水下进行探秘，全世界的地质学家、古生物专家纷沓而至帽天山考察，在此拉开了进军未知世界的序幕。

抚仙湖虫，这个名字首先问世于1987年《古生物学报》26卷第三期。侯先光教授在题为《云南澄江帽天山早寒武世三个新的大型节肢动物化石》一文中，阐述了1984年新发现的三个新物种，并且将其中的一个新物种命名为“延长抚仙湖虫”，拉丁文为 *Fuxianhuia protensa*。解铃还须系铃人。2006年11月6日，我就这道怪题拜访了抚仙湖虫的命名者侯教授。

原来，我们居住的地球是太阳系中的一颗行星。根据地球上最为古老的岩石测定，地球年龄大约有46亿年，可以将它划分为冥古宙、太古宙、元古宙和显生宙四个演化阶段。冥古宙时期（46亿至38亿年前），地球多了一个月亮妹妹。45亿年前，地球与一颗星体斜向相碰，碰出月亮的诞生和地球的熔融。地球的熔融使密度大的铁、镍等元素沉积于内核，使密度小的岩石聚集于表层而冷却为地壳。太古宙时期（38 亿至 25 亿年前），生命细菌问世于地球，但还处于原核细胞状态。元古宙时期（25 亿至 5.4 亿年前），新生物诞生于地球。大约 16.5 亿至 14 亿年前，大气层中的含氧量是现在的 1%，厌氧的原核生物衰退，喜氧的真核细胞诞生。由于罗迪尼亚泛大陆分裂而诱发大规模火山，导致地球降温与大冰期（雪球事件），大气层的含氧量达到现在的 7%，真核生物挣扎，新生命孕育。元古宙晚期（5.4 亿年前），是动物出现与演化的时代。扬子次大陆远离罗迪尼亚泛大陆，而且成为沉积了生息的家园。此时至显生宙早期大气层的含氧量达到现在的 10% 左右，碳酸岩、砂岩和含磷地层已经形成，具备了动物体分泌矿化骨骼的条件，于是澄江帽天山寒武纪生命大爆发。

地球历史悠久，生命演化漫长，生物化石如凤毛麟角。澄江帽天山动物化石群是人类首次看到的最古老、最完整的海洋生物群落图。侯教授于 1984 年至 1985 年发现的动物化石之一：体长超过 10 厘米；头甲如半圆，头顶有一对复眼和一对触须，头下有口板；身子与尾巴由31个体节组成，体节上长有腿肢，犹如一个精美的甲壳动物。这枚化石是世界上最原始的节肢动物化石，是节肢动物的祖先。它在世界科学领域具有广泛的影响，如何赋予它一个名副其实的名称？这比父母为孩子取名字的责任更重大！

《广雅·释亲》云："姓，子也。"王念孙《广雅疏证》载："姓者，生也。"我国公民的姓名，既标志着家族系统的称呼，又要避免重复而好听吉祥，有的甚至承载着父母的寄托，寄寓着文化的内涵。生物化石命名必须根据国际命名法规：一个新的物种名称不但应适合现代生物，也应适合古代生物。物种的命名采用双名法则，即一个物种的名称必须由两个词组成，第一个词为属名，第二个词为种本名。属名是生物命名中最为重要的名称，也是种本名依附的支柱，犹如人的名字必须和姓连在一起，才能构成一个完整的姓名。根据命名规则，属名和种本名都可以用化石的分布范围和产地命名，观其名即可知其发现区域和发现地点。侯教授看着这枚世界级的重要化石，首先联想到的是世界著名的抚仙湖：两百多亿立方米纯净可口的一类淡水，清澈透明，位居我国淡水湖泊前列。澄江帽天山动物化石群位于抚仙湖的东北岸，因此，侯教授将这枚世界级的重要动物化石命名为“延长抚仙湖虫”。加“延长”二字，指该动物体形较长。它寄意着，让古老的动物化石像美丽的抚仙湖一样名扬世界，让美丽的抚仙湖像古老的动物化石一样保存着地球演变的故事。

我听到此处，情不自禁地赞赏道：“侯教授，这可作为化石命名的典范！”他却十分谦虚地说：“我长期在抚仙湖畔工作，情理所在。”是啊！“情理”二字震惊世界。

澄江化石地展览馆

抚仙湖虫于《古生物学报》问世后，美丽的抚仙湖与帽天山古老的动物化石备受国内外游客和专家、学者青睐。

侯教授越讲兴趣越浓，用纸笺将那三个大型节肢动物的名称写给我，并且提议由我为抚仙湖虫题诗留念。诗，要意象，讲寄意。无七步之功，难以奏效。我为了不让侯教授失望，勉强口占《抚仙湖虫》二首：

众生多落没，
遇主得成名。

澄江化石地展览馆中心展厅

莫怨苍茫地，
从来路不平。

名扬真不易，
古董作文明。
倾吐衷肠事，
人生也变更。

小诗虽然不算好，但基本符合韵律，多少起到了一点助兴的作用。侯教授笑得眯起双眼，乐滋滋地投入授业解惑：“距今 300 多万年前的第三纪末期，由于地壳板块急剧运动，造成喜马拉雅山脉凸突，从而引起一系列断层贮水及石岩溶蚀现象，在云南高原群山中形成了众多的湖泊。抚仙湖就是此时形成的一个南北向的断层陷落湖。它比 5 亿多年前的抚仙湖虫年龄小得多啦！抚仙湖虫应该算抚仙湖的老祖宗。这比喻不一定恰当，但可以说明其诞生的时间差异。”

我与侯教授交谈三个多小时，犹如遨游乾坤，历览天地。一枚化石就像一部伟大的史诗，给人以智慧，给人以力量，给人以洞察深远问题的眼力。古代圣贤，爱登高望远，抒怀壮志，以古论今，或“登泰山而小天下”，或“独上高楼，望尽天涯路”，或“念天地之悠悠，独怆然而涕下”。今朝志士，喜读石深思，寻规悟道，以今说古。达尔文从化石标本里读出了“选择演化论”，列文虎克将一块石头磨成世界上第一个显微镜，侯先光从化石里读出了生命摇篮。人生如是，善于穷尽乾坤，有看法便能看得透，有想法便能想得通，有说法严格遵纪守法。

我们的这次交谈，结束在一个充满诗意的环境中：碧水清清云淡淡，仙湖虫语鸟欢声。

抚仙湖虫复原

Fuxianhuia

When Prof. Hou Xianguang found the Maotianshan Fossil Site, he named one kind of species *Fuxianhuia*. This caused the curious into speculation: What came first? *Fuxianhuia* or the Fuxian Lake? It poses a chicken-and-egg conundrum for astronomers, long curious about which came first. Perhaps to find out answers to this mysterious question, China Central Television (CCTV) visited Fuxian Lake twice to shoot of underwater world, exploring secrets. Geologists and paleontologists from around the world have been going to Maotianshan Mountain for investigation. They are here beginning to confront the world of the unknown.

Fuxianhuia was firstly published on "Early Cambrian large bivalved arthropods from Chengjiang, Eastern Yunnan" in the journal of *Acta Palaeontologica Sinica* 26(3), 1987. Professor Hou Xianguang reported three new species which were found in 1984. One of the new species was *Fuxianhuia* protensa. In order to untie the bell, the person who tied it is required. I visited the Prof. Hou who first described *Fuxianhuia* on November 6th, 2006.

The earth we inhabit is a planet in the solar system. Based on the method for determining the age of the oldest igneous rocks, the age of the earth is about 4.6 billion years, which can be divided into Hadean, Archaean, Proterozoic and Phanerozoic eons. During Hadean eon (4.6 – 3.8 billion years ago), the moon was formed. The moon is thought to have formed from debris ejected into space after a Mars-sized body collided with the still-molten Earth about 4.5 billion years ago, probably with an iron/nickel core and a silicon mantle. In Archaean eon (3.8 – 2.5 billion years ago), bacteria in prokaryotic cell came out on the Earth. In Proterozoic eon (2.5 billion – 540 million years ago), new creatures were birthed on the Earth. About

1.65 – 1.4 billion years ago, the oxygen content in the Earth's atmosphere is one percent of today. Anaerobic prokaryote had a recession and aerophile eukaryocyte were present. The breakup of Rodinia caused the massive volcanic, then came several glacial events around the world which were called "Snowball Earth". By that time, the layer was with an oxygen density probably about 7 percent. When eukaryotes were during the struggle, there was new life breeding. Late Proterozoic (540 million years ago) was an age of animals appearing and evolution. Since the Yangtze Block was not part of Rodinia, it became a homeland to support life. At that time to Early Phanerozoic, oxygen content of the atmosphere stood at about 10 percent of today. Carbonate rocks, sand rocks and Phosphorus formation had formed, which were necessary conditions for secretion of mineralized skeletons of animals. Then it came out the Chengjiang Maotianshan Shale in the "Cambrian Explosion".

The earth has a long historical standing and so does the evolution of life, but the fossils are very rare because of the preservation conditions. Chengjiang Fossil Site on Maotianshan Mountain is so far the earliest and the most completed biocenosis in the ocean. *Fuxianhuia* was a species found by Prof. Hou in 1984 – 1985: Body length more than 10 cm; the anterior body encased in an oval sclerite, from which two stalked eyes emerge; paired antennae; oral under the head shield. There are 31 limb-bearing segments comprising the thorax and the tail of *Fuxianhuia*, looking as if an exquisite crustacean. *Fuxianhuia* is the most primitive fossil arthropod in the world, of which the primitive features have led to its playing a pivotal role in discussions about the arthropod stem group. *Fuxianhuia* has wide-ranging implications in the world sciences research. How to give it a name worthy of it? It's a greater responsibility than new parents choosing name for their children.

As *Guangya·Shiqin* saying, "The family name is for the descendants." Wang Niansun wrote in *Guangya Shuzheng*, "The family name is to carry on." In our country, family name and given name not only represent the family system, but have "the great auspicious"meaning. Parents put their heart to give children the name that is distinctive and cultured. To name the fossils must be in accordance with International Code of Nomenclature. A new species must have the name suited for modern creatures and ancient beings. Biologists and paleontologists use the binomial system of nomenclature developed by Linnaeus to categorize the varieties of life on the earth. A species should have at least two names — a genus name, which is general and capitalized, and a lower-case species name that is more specific. Genus name is the most important name like the family name. A completed name must contain the genus name and species name, which are often after a distribution range or a place of origin. Prof. Hou connected the world-famous Fuxian Lake with the world-famous fossils. Fuxian Lake contains more than 20 billion cubic meters of fresh water, clean and transparent. It is on top of the freshwater-lake list. Chengjiang Fossil Site on Maotianshan Mountain lies on the northeast bank of Fuxian Lake. Therefore, Prof. Hou named this world-class important fossil *Fuxianhuia protensa*. *Protensa* implies its protensive body plan. Moral of the name: Let the ancient fossils become famous all around the world like the beautiful Fuxian Lake; let the stunning Fuxian Lake tells the stories of the evolution of the earth like the oldest fossils.

When hearing this, I couldn't help saying, "Prof. Hou, this can be served as an exemplary manner to name fossils." He said very modestly, "I've been working by the Fuxian Lake for a long time. That's the emotive reason." The emotive reason, of course! These two words are earth-shattering. After the article getting published in *Acta Palaeontologica Sinica*, oldest fossil *Fuxianhuia* from Maotianshan Mountain and beautiful Fuxian Lake are so heavily favored by domestic and foreign travelers, experts and scholars.

The more Prof. Hou talked, the more infected he became. He wrote the names of the three large bivalved arthropods on a piece of paper, and suggested, "Would you write some poems of *Fuxianhuia*?" Poem embodies the beauty of artistic conception. Without the talent of "Seven Steps into Poetry", it makes poetry difficult to write. I made two poems about *Fuxianhuia* for Prof. Hou:

Most of creatures go extinct,

Meeting a talented person is why they became legends.

Don't blame the boundless land,

There is always rough road.

It's not easy to come to fame,

The oldest objects are the most potent civilizing forces.

By breathing out the things of soul,

The life they have shared is changing forever.

These little poems didn't contain brilliant metaphors, but at least they could rhythm. The rhythms made some help for our entertainment. Prof. Hou began to impart knowledge and explain the puzzles, smilingly. "At the end of the Tertiary period, 3 million years from now, a rapid uplift happened in Himalaya Mountains involving with plate tectonics, which caused a series of faults and karst topography. In those mountains on the plateau of Yunnan Province, numerous lakes were formed. The Fuxian Lake, a north-south depression lake due to the collapse, was born that way. Fuxian Lake was much younger than the 500-million-year-old *Fuxianhuia*! *Fuxianhuia* is like the forefather of Fuxian Lake. This metaphor might be kind of inappropriate, but it could tell the time difference of their birth."

I've talked with Prof. Hou for over 3 hours. That conversation was without borders like a heavenly steed soaring across the skies. The fossil specimen may represent like a great epic to give people wisdom, power and a view inside to profound questions. The ancient sages liked looking out over the mountains and poured out their hearts and aspirations, which took current thought as that of ancient. They either felt "Lu kingdom is smaller when one is on the top of the East Hill, and the world is smaller when one is on the top of Tai Mountain", "Climbing up the stairs and being lonely on the loft, one overlooked the endless distance", or "Thinking of heaven and earth, without limit, without end, and one felt all alone and tears fell down". Nowadays, the ambitious people like studying with stones in search of the truth with a thoughtful mood. They make the research of ancient world from the experience of modern times. Darwin concluded the theory of evolution by natural selection from the facts of the fossil specimens. Leeuwenhoek invented the first electronic microscope by sculpting a lens. Hou Xianguang found the cradle of life from the fossils. That's what life is about — people can generate interesting ideas from minor things and once the ideas are formed, they have strong urge to put it into action.

Our talk ended in a poetic atmosphere. The Fuxian Lake is clean and the cloud is light, with insects whispering and birds singing.

奇虾化石

话说奇虾化石真个奇！中国人用它做邮票，美国人用它做广告，日本人也紧跟春潮枝头闹。

奇虾化石邮票

奇虾邮票用的是奇虾的软体组织印痕化石照片，橙黄色的岩石上，清晰可见奇虾软体的棕褐色印痕。这是1999年中国科学院建院50周年时，中国邮电部发行的4枚具有世界科技影响的特种邮票之一。它向世界昭示，华夏文明古国继发明造纸术、印刷术、指南针、火药之后，又有了四大科技文明成果，即：以奇虾为代表的澄江化石、智能工程的标志“6000米水下机器人”、理论数学的桂冠“哥德巴赫猜想”、天文学的成就2.16米天文望远镜。科学家认为：奇虾类动物软体化石是最珍贵的化石之一。动物死亡腐化了，一般无任何痕迹，假若它有万分之一的可能侥幸形成化石，那么也只可能是坚硬的牙齿、骨骼或外壳得以矿化。预想不到的奇迹是：这枚“寒武纪早期澄江生物群”的奇虾化石邮票，是一个完整的奇虾矿化图像。它标志着澄江古生物化石群在当今世界古生物学研究方面的高层次水平。

奇虾广告用的是三维动画，形象威武俊美：流线型的身上披挂着庞然甲衣，尾翼飘逸；双臂长满锐刺，左右锋利夹击；嘴巴巨大，牙齿像刀片一样排列，可以将食物肢解为小块，嚼成肉末；一对带柄的巨大眼睛，像探照灯一样追射着原始海洋里的一切猎物；背腹扁平，善于游泳；两米多的长身躯上焕发着翻江倒海的伟力。当时的原始海洋里，一般动物的身躯只有数毫米或数厘米长，奇虾凭借着庞大的体型和无与伦比的精良装

2006 年 8 月 12 日曾培炎视察澄江古生物化石地

备，任意掠食，所向无敌，称王称霸。它就是地球上的第一个掠食者。从此，暴力诞生，战争兴起，终结了由简单的海绵动物统治海洋的时代，开始了从原始低等生物向现代高等复杂生物进化的时代。掠食者与被掠食者之间的生存竞争愈演愈烈，并且延续至今天。不过，奇虾首创竞争先例的环境条件是地球运动的结果。地球运转无止境，动物生存终有限，任何动物也难逃脱这一规律。奇虾，肆无忌惮地生存了 900 万年之后，终于在地球的伟大演变中灭绝了！

奇虾真的灭绝了吗？我们作为单个生命的每一个人，最多活上百把岁就死亡了，如果作为种族生命的人类，却生生不息！以此推测：奇虾种族可能已经被新的生命所代替。

奇虾复原图

这尚待科学家揭秘。

奇虾种族生命延续的揭秘，虽然还是一个未知数，但是奇虾化石问世就像一个具有伟大思想艺术的人一样，他的生命虽然消逝了，但是他的思想文化生命却生生不息。我们本家祖宗孔子的生命虽然早已融入大地，但是他的伟大思想仍然活脱脱地影响着人类世界。奇虾化石虽然没有思想，但是它成为东西方文化认可的共同“偶像”，这也许是它为了生存而开创了竞争！君不见：奇虾问世五亿三千万年，江河海洋里大鱼吃小鱼，适者生存；山林草地上猛兽掠食弱者，物竞天择。君不见：人猿相揖几千年，风流人物挽弓满月，捍卫正义；名利场上锱铢必较，取之有道。人类以生存竞争为核心的进化，从无序走向有序，从野蛮走向文明。人，是由原始脊椎动物逐步进化而来的，也是由各种物质细胞组合而成的，如果只知道争夺豪奢，就与动物无异。为什么有的人，生活愈富裕愈冷酷无情？本性使之矣！人类，靠修身重品德逐步摆脱了愚昧状态，以文明进步区别于单一追求物质享受的动物。正确的人生观，也许来自于科学的宇宙观；科学的宇

2003 年陈均远（左三）教授与徐荣凯（右二）合影

宙观，也许来自于运用地球演变和动物进化的研究成果。

什么样的地球环境产生什么样的动物！什么样的土壤生长什么样的种子！什么样的制度孕育什么样的人才！1984 年，侯先光教授在澄江帽天山发现动物软体组织化石之后，陈均远教授和周桂琴女士于 1992 年在澄江动物化石群发现了完整的奇虾化石标本，舒德干院士等也在这里发现了一批动物软体组织化石。近 30 年来，国内外的科学家在澄江动物化石群采集了10万件古生物化石，影响着不少科学家循着遥远祖先的足迹，研究和思考生命历史的方式。这些善于悟透古生物化石的科学家中，也许将会出现哥白尼、达尔文似的伟大天才！这是我的期盼，也是一个“赌咒式”的预言：

九百万年称霸王，
一朝巨变倾巢亡。
帽天山顶英灵在，
给力酬勤誉满堂。

Anomalocaris

The fossils of *Anomalocaris* are really surprising! Chinese people printed them into the stamps and American people created advertisements by them. *Anomalocaris* is also very popular in Japan.

Anomalocaris stamp uses a picture of impression fossil of soft tissue. From the picture, you can see clearly brown impression of *Anomalocaris* soft tissue on the orange rock. It was in 1999, the 50th Anniversary of the founding of CAS, that PRC Ministry issued 4 special stamps with global scientific and technological influence. *Anomalocaris* stamp is one of the 4 stamps. It declares to all over the world that after inventing paper, printing, compass and gunpowder, the Chinese ancient civilization research and develop scientific and civilized achievements in 4 aspects, which are Chengjiang fossil taking *Anomalocaris* as representative, the sign of intelligent engineering — underwater robot 6000 meters below the surface, the laurel of pure mathematics — Goldbach's conjecture, the astronomical achievement 2.16m astronomical telescope. Scientists assume soft tissue fossil of *Anomalocaris* kind is one of the most precious fossils. Normally one animal leaves no mark after its death and decay. If it turns to fossil in a chance in a million, it can only be mineralized tooth, bone or skeleton. But there is an unexpected miracle that this stamp of *Anomalocaris* fossil, which belongs to the Chengjiang Biota of early Cambrian, is a complete *Anomalocaris* mineralized image. It indicates Chengjiang Biota is the highest level that paleontological research can achieve in the world.

The *Anomalocaris* advertisement is based on three-dimensional animation. From it you can see: the image is powerful and pretty; the clipper-built body wears a mighty armor coat, with a flowing empennage; both arms covered with sharp spines are like sharp attacking from both sides; the big mouth has teeth standing like blades, which can dismember food to small pieces and then chew it into minced meat; a pair of large stalked eyes are just like searchlight, shining on and shooting prey in the primitive ocean; the flat back is good at swimming; the over 2-meter long body glows with stormy power. Common animals only have several millimeters or centimeters bodies in the primitive ocean at that time. But *Anomalocaris* is able to catch food easily and invincibly, like a king, relying on its huge body, incomparable and superior equipment. It can be said to be the first predator, as its appearance gave birth to violence and war, it ended the era of simple sponges manipulating the ocean, and started the new page of evolving from lower-level primitive creature to modern higher-level complex creature. The struggle for life between predators and prey is becoming even more violent and extends to nowadays. However, the environmental conditions which enable

Anomalocaris to set a precedent of competing gives the credit to the movement of the earth. The turning around of the earth is everlasting, while the survival of animals is limited. Any animal is unable to violate this rule, including *Anomalocaris*. So even it has lived over 9 million years unscrupulously, it finally dies out during the great evolution of the earth.

But does *Anomalocaris* really die out? Each personal as a single life can live over 100 years at most, but human beings as racial life are able to live circle of life. Depending on this speculation, *Anomalocaris* may be replaced by brand-new creature, which needs the scientists to discover and answer.

It is still unknown to discover and answer the extending of *Anomalocaris*' racial life. The coming out of *Anomalocaris* fossil is similar to the birth of a human being with significant thinking and art. Although his life came to death, but his thinking and art would grow over and over again. The life of our ancestry Confucius blended into with the earth, but his thinking is still affecting the human world very significantly. Although *Anomalocaris* fossil has no thought, it has become a common icon approved by eastern and western culture, maybe because it begins competition for survival. Have you ever seen that in 530 million years since the appearance of *Anomalocaris*, big fish eats small fish in the sea and rivers, only the strongest survive; the predators catch and feed on the prey in the mountains and forests, which is survival of the fittest in natural selection. Have you ever seen that the species of man ape live peacefully and respectfully for thousands of years, but only truly great men defend for justice; in Vanity Fair, fight over the smallest trifles, and make money in proper ways? Human beings evolve around the core of survival and competence, from disorder to orderly, from barbarous to civilize. Human beings are evolving gradually from primitive vertebrates and consist of various substance cells. If human beings only fight for luxury, then they have no difference from animals. Why does somebody become more merciless when he gets richer? Maybe it's owing to human nature. Human beings get rid of ignorance because of self-cultivation and shaping morality, as well as from civilization and progress differ from animals going after material comfort. Correct outlook on life possibly comes from scientific world view, while scientific world view is from research achievement on the evolution of the earth and animals.

The type of animal depends on which kind of earth environment; seed depends on earth, talents on system. In 1984, Professor Hou Xianguang found specimens with soft tissues in Maotianshan Mountain of Chengjiang. After that, many scientists have found a new batch of soft tissue fossils there. During almost 30 years, the scientists at home and abroad collect 100,000 pieces of paleontological fossils there. All of these make many scientists follow trace of deep ancestry and influence the way of their researching and thinking of the life history. Maybe a genius like Copernicus, Darwin will come out among these scientists adapt in paleontological fossils thoroughly. This is my expectation, as well as swearing prediction as below:

Claim as the lord and king in 9 million years,

Lose the castle and die out after overnight change.

Spirit of an esteemed person exists in Maotianshan Mountain,

God rewards the diligent with fruitful results and honor.

华夏鱼化石

话说人从何处来？先有男人，还是先有女人？

中国神话传说里或讲人是女娲造的，或讲人是从葫芦里生出来的，教科书里讲人是猿类进化的。那么猿又从何而来？西方神创论者认为："地球上的万物是在公元前4004年10月26日9点，由上帝创造出来的。"如此众云，影响了人类社会几千年。公元1543年，哥白尼的《天体运行论》诞生，神学自然观开始受到真实宇宙观的冲击，人类知道了天外还有天：地球围绕太阳转。1859年，达尔文的《物种起源》诞生，人类知道了生命的进化过程与法则。但是，达尔文还希望在寒武纪化石中找到有力的证据，这个

2012年11月于北京，舒德干院士向作者赠送译著《物种起源》

作者与杨兴荣、李朝伟研究帽天山规划

伟大的愿望在云南澄江帽天山实现了！

侯先光开创帽天山古生物化石地十载之后，舒德干一行在距离帽天山大约一公里的小滥田村附近，从无数浅黄色页岩中敲出一块神秘的化石。这块页岩面上沉睡着一条“V”字形的小鳗鱼：背部有一条深深的褶皱痕，还有肌肉节、脊索印痕和鳃裂，长 2.2 厘米。舒德干与剑桥大学康威·莫里斯、张兴亮合作，经过一年多的研究，将这件 5.3 亿年前的小鳗鱼化石命名为“华夏鱼”，并且于 1996 年 11 月 14 日在《自然》杂志上向世界公布：华夏鱼是地球上最早的原始脊索动物，是包含人类在内的脊椎动物的最早祖先。十年磨一剑。舒德干院士等曾经在澄江化石地发现世界最古老的脊椎动物昆明鱼化石，将包括人类在内的脊椎动物的起源提前了五千年，被称为20世纪国际古生物界最重要的研究成果之一。从节肢动物到其他各种类型无脊椎动物的发现，再到脊椎动物的发现，证实澄江化石地是人类生命的摇篮！

生命的本源，发祥于帽天山！

这个判断，是建立在千千万万个古生物化石的研究基础之上的结论。纳罗虫、三叶虫、抚仙湖虫、奇虾、华夏鱼是帽天山发祥地的标志性古生物化石。这五件标志性化石的发现和研究者，是无数科学家中的杰出代表。他们希望从澄江化石宝库里窥视并解绎出寒武纪生命大爆发的一些奥秘，希望破解达尔文关于“人类很可能源自低等动物”的猜想，并努力为之提供由低等动物演化到人类的真正历史证据。无数学者对澄江化石宝库深入探索与发现的这五件标志性古生物化石，就是对这一历史进程的最可靠证据。

这些最可靠的证据，震撼了古老的世界！

1992年10月，罗马教皇约翰·保罗二世宣布伽利略《关于托勒密和哥白尼两大世界体系的对话》一书不再是“异端邪教”了，从而为这位蒙冤三个半世纪的意大利伟大科学家“正式平反”。1996年，在生物进化论日趋深入人心的大背景下，这位教皇不得不改变教会对达尔文主义的否定态度，他致函教廷科学院全体会议：“（天主教的）信仰并不反对生物进化论”，“新知识使人承认，进化论不仅仅是一种假设”，“事实上，由于各学科的一系列发现，这一理论已被科学家普遍接受”。澄江化石地发掘的一系列古生物化石，进一步夯实了这两个被平反的科学基石。

人类逐步深化着对宇宙的认识：从敬畏大自然、崇拜大自然的宗教文化，转向敬重大自然、顺应大自然的科技文明。这个转折点始于哥白尼，加速于达尔文，成熟于帽天山古生物化石地的问世。人类的原始宗教文化，为约束人的非正当行为起到了历史性的作用，同时也限制着人的个性发展。科技文明解放了人的个性，激励出信息网络与卫星上天，助推着人类从神在天上转向神在自己心中！

人生有涯而知无涯。一个人如果天天不停地读书，即使一百年也读不完当今世界的全部的出版物。许多书其实可以不读，但是唯有帽天山这部无字之书不能不读！因为追溯历史：野蛮人类到文明社会，只动用了三个苹果；猿猴到人类，只有几个石头轻轻擦过；寒武纪大爆发，却是3000万年的演化；地球诞生46亿年，萌生了这几件古生物化石。从此，揭示了从低等动物到高等动物、从简单到复杂的自然演化过程。这个演化的机制便是：适者生存，自然选择。

人类怎样正确地生存？看来只有运用正确的宇宙观，才可能有正确的生产生活方式。这就是帽天山的价值。

Cathaymyrus

Question: Where did the human come from? Who came first, the male or the female?

In Chinese myth, humans are created either by the goddess Nu Wa or from a hollowed gourd. In the textbooks, humans evolve from apes. Where do apes evolve? The western creationists hold, "at 9:00 on October 26th about 4004 B.C., all creatures on the earth were created by God." This theory has affected the human society for thousands of years. In 1543, Copernicus's *De Revolutionibus* impacted the theology view of nature, making people realize the earth goes around the sun. In 1859, Darwin's *Origin of Species* made people aware of the evolution of life. However, Darwin's expectation to find out more convincible evidence in the Cambrian fossils was accomplished in Yunnan

Maotianshan Mountain.

10 years after Hou Xianguang's discovery, Shu Degan and his colleagues found a mysterious and incomplete fossil from the yellow slabs around Xiaolantian Village, which is about 1 km away from Maotianshan Mountain. This only specimen is a V-shaped little fish of 2 cm length: The dorsal part of this animal preserved a deep muscle scar, V-shaped muscle blocks, notochord trace and gill pouches. Shu Degan, cooperating with Conway Morris from the University of Cambridge and his colleague Zhang Xingliang, after more than a year's study, published on *Nature* journal and named this 530-million-year-old fish *Cathaymyrus* on November 14, 1996. They announced to the world: *Cathaymyrus* is the earliest primitive chordate on earth, which is also the earliest ancestor of human vertebrates. Decade of grinding sword, Shu Degan, academician of Chinese Academy of Science, used to find the earliest chordate *Myllokunmingia fengjiaoa* in Chengjiang Fossil Site. This moved the origins of vertebrate

human life up by 5,000 years and was called “One of the most significant paleontological researches in the 20th century”. The discoveries from arthropod, other invertebrate to vertebrate prove that Maotianshan Mountain is the cradle of human.

Humans originate from Maotianshan Mountain!

This judgment is based on thousands of researches on fossil specimens. Specimens of *Naraoia*, trilobite, *Fuxianhuia*, *Anomalocaris* and *Cathaymyrus* are landmarks standing as symbols of Maotianshan Shale. The discoverers and researchers of them are outstanding advocates for numerous scientists. They wish to find out and explain the mysteries of Cambrian Explosion from Maotianshan Mountain. They want to decipher Darwin’s guess that humans may originate from lower animals, and strive to provide some real historical evidence. In fact, the five symbolic fossil specimens are the most convincible evidence.

All these evidence shock the ancient world!

In October 1992, Pope John Paul II announced Galileo's *Dialogue about Ptolemy and Copernicus Two World System* was not heresy anymore, officially rehabilitating the great Italian scientist wronged for three and a half centuries. In 1996, when evolution won support among people, the pope had to change the church's negative attitude toward Darwinism. He sent a letter to Academy of sciences, the plenary session, saying that, "Catholicism doesn't object evolution", "new knowledge makes people admit evolution is not just assumption", "in fact, a series of discoveries have been accepted by scientists". The fossil specimens unearthed in Maotianshan Mountain further consolidate the scientific basis.

Human's understanding of the universe gradually deepens: from the religious culture of admiring and worshiping the nature to the scientific culture of respecting and conforming to the nature. This turning point starts from Copernicus, accelerates from Darwin, and matures from fossil specimens unearthed in Maotianshan Mountain. The original religious culture plays a historic role in restricting the indecent behavior, but restraints the individual development. The scientific culture emancipates human's individuality, stimulates Internet and satellite, and convinces people to believe god in heart instead of god in heaven.

Life has limits, but knowledge has no limits. Even if one reads non-stop for one hundred years, he can't read all the publications in the world. Some books can be neglected, but the intangible book Maotianshan Mountain is a must. Tracing back history, we realize that three apples change the world: some stones rub and make apes evolve into humans; Cambrian Explosion lasts 30 million years; 4.6 billion years only produce these fossil specimens. From then on, the natural evolution from lower animals to higher ones, from simple to complicated has been unveiled. The principle is the fittest survive and nature selects.

How can the human race sustain correctly? The right mode of production and living is possible only with a correct view on the universe. This is the value of Maotianshan Mountain.

老故事

话说玩石头玩到联合国的故事。从帝王将相玩珠宝玉石，到平民百姓玩观赏奇石，玩石头者多如繁星，谁把石头玩进了联合国？

圣彼得堡斯莫尔尼宫的坐标为：东经 30°25'，北纬 59°55'。联合国第 36 届世界遗产大会在此举行。公元 2012 年 7 月 1 日（北京时间）22 点 12 分，此次大会开始审议澄江化石地项目。这是杨洋副市长从莫尔尼宫给我发来的手机信息。我们当时所在的澄江化石地的坐标是：东经 102°58'38''，北纬 24°40'08''，两地时差整整 4 小时。公元 2012 年 7 月 1 日（北京时间）22 点 22 分，中国云南玉溪澄江化石地正式列入世界遗产目录。

此时，澄江县古生物化石馆前的广场上空，星光灿烂，电波频传。星光下参加庆祝活动的群众载歌载舞，各大媒体记者奔波于申报审议会场与此地的实况直播工作。我代表玉溪市致辞：“澄海上，万物生，好梦变成真！昨天的好梦，此刻已经实现。明天的好梦，必将变为事实！”好梦变成真，是我创作的《生命摇篮》歌词的苦吟之句，也是醉吟警句。小小化石梦，百年漫长，趣味无穷。古老的故事有必要在今天和盘托出，虽然前面曾经有过点睛之笔。

法国进入资本主义上升时期，疯狂地向外扩张势力和掠夺资源，并派遣一批高素质的自然科学家到世界各地进行科学考察和军事探险活动。1904 年 7 月，奉命赴任印度支那地质调查所所长的德普瑞特 J. Deprat，1880 年生）和满苏（H. Mansuy，1857 年生），曾经多次到中国云南澄江和越南、柬埔寨等地进行地质考察。1907 年至 1912 年，他俩合作发表、出版了一些关于云南地质考察研究的论文和著作，其中 1909 年出版的《滇东地质状况备忘录》首次向世界公布：澄江帽天山属于寒武纪地层，并发现寒武纪古生物化石。遗憾的是，这一伟大的发现，却被奸佞小人诬陷为“编造化石”的科学欺骗事件，导致优秀的地质科学家德普瑞特失去了印度支那地质调查所所长职务。更遗憾的是，他们已经没有条件去发现澄江动物化石群，一切追寻都变成了梦幻。后来，道格拉斯·帕尔摩撰文《布尔吉斯页岩的一个对手》，高度赞扬了这两位优秀的地质学家，并且悲壮地揭示了地质科学史上的曲折艰辛。

1939 年初，中山大学数千名师生逃脱日本侵略军的屠杀，将校址搬迁到云南澄江县，中山大学理学院地质系即安置在澄江跨马村里。1940 年 3 月，著名地质学家何春荪与鲍礼仪教授赴此校任教，并深入澄江东山调查磷矿资源。1942 年，何春荪教授在《地质汇报》第 35 号上发表《云南澄江东山磷矿地质》的文章，首次向世界公布帽天山地质图和剖面图，并且提供了帽天山“页岩内有一种低等生物化石”和“德国米士教授获有三叶虫化石”等依据。何春荪教授望重资深，早年就职于国民政府中央地质调查所，1943 年奉命与毕庆昌等赴兰州组建中央地质调查所西北分所；1949 年到台湾大学地质系任教。这位发掘帽天山古生物化石的奠基者，是否阅读过发掘帽天山古生物化石的始祖德普瑞特的论文，我至今尚未查询到可靠的证据。不过，第一个发现帽天山古生物化石群的著名地质学家侯先光，却是仔细地研读了他们的原始资料，而且尾随他们的足迹，从南京到帽天山，于 1984 年 7 月 1 日完成了他们未尽的伟业，被世界科学界誉为 20 世纪最惊人的发现之一。

从此，澄江化石地成为国内外瞩目的宝地。1987 年，澄江县公布《关于保护澄江无脊椎动物化石群的暂行规定》，成立澄江动物化石群保护管理机构。1992 年，联合国教科文组织将澄江化石地列入《全球地质遗址名录》。1997 年，云南省政府将 18 公里的部分澄江化石地划定为自然保护区。2004 年 7 月 1 日，原云南省省长和志强赴帽天山考察。他是地矿工程师，并首次提出“向联合国申报世界自然遗产”的构想。玉溪市县两级旋即组织推进“申遗”筹备工作。2004 年 9 月 5 日，温家宝在新华社内参上批示：“要保护澄江化石群，保护世界化石宝库，保护这个极具科学价值的自然遗产。”云南省省长徐荣凯批示：“坚决落实温总理批示要求，采取有力措施，保护澄江化石群。”

玉溪市立即召开市县党委常委会议，研究部署落实温总理和徐省长的批示精神：含

玉溪市贯彻落实温总理“三个保护”现场办公会

2016年玉溪市长饶南湖与澄江县委书记李朝伟在帽天山调研

着眼泪关闭化石地周边矿山，共计14个磷矿点价值167亿元；勒紧裤袋投入保护资金近亿元，退耕植树2万多亩；咬紧牙关取缔油污抚仙湖的1000多条机动船艇，并借此机会加大世界自然遗产申报工作，确定市长助理强卫东负责具体事宜。2005年11月，云南省人民政府向国务院呈报《澄江化石地申报世界自然遗产的请示》。2006年，中国政府向联合国推荐将澄江化石地列入《世界自然遗产预备名录》。2007年，国务院副总理曾培炎与中共云南省委、省政府领导考察澄江化石地，对保护世界化石宝库和“申遗”工作提出了更高的要求。2009年4月，玉溪市委、市政府确定杨洋副市长负责推进澄江化石地“申遗”事宜。

2011年，中国政府向联合国提交澄江化石地申报世界自然遗产材料，联合国派专家组赴澄江化石地考察评估。2012年6月24日（当地时间）17时，杨洋副市长率领澄江化石地“申遗”小组到达圣彼得堡莫尔尼宫，鼎力做圆这个历史性的好梦。

此情历历成追忆，世事艰辛出奇迹！

从2012年7月1日追溯到2004年7月1日的8年间，是全力保护澄江化石地的重要时期，积累了许多新的经验，启示着我们必须建立一个与世界接轨的澄江化石地保护管理机构，承担起历史性的重任。

从2012年7月1日追溯到1984年7月1日的28年间，是以侯先光、陈均远、舒德干为代表的无数杰出科学家科考研究硕果累累的辉煌时期，打造出了这座地球生命摇篮的不朽品牌，启示我们必须建立一个服务于国内外科学家的科研平台，借天下之才发掘澄江化石地的科学价值。

从2012年7月1日追溯到1904年7月

的108年间，是德普瑞特、满苏、何春荪等若干科学家探索发现澄江化石地的奠基时期，留下了悲壮的惊世故事，启示我们必须建立一个能够召开世界地质年会的展示馆，向世界展示这一伟大的奇迹，普及科学的宇宙观。

石缘好梦百年期，喜讯巡天庆“申遗”。今后的故事内容也许是：追往事，梦明天，期盼“申遗”东风吹醒化石，古代生灵重游大千。

2012年7月遗产大会会场

申遗成功，玉溪代表团“沸腾”了

申遗成功，玉溪市副市长杨洋接受祝贺

①

②

③

申遗代表化石

①迷人林乔利虫

②中华微网虫

③铅色云南虫

④相连群体行为虫

⑤耳材村海口鱼

⑥澄江东龙潭贝

⑦长足罗哩山虫

Old Stories

I would like to tell a story about playing with stones at the United Nations. Emperors and nobles play with the jewel and jade. Civilians play with ornamental and grotesque stones. The stone-collectors were numerous as stars, but who brought the stones to the United Nations?

Coordinates of St Petersburg Smolny is 59°55' north latitude, 30°25' east longitude. The thirty-sixth session of the United Nations World Heritage conference was held here. At the time of 22:12 on July 1st, 2012 (Beijing time), the conference began its consideration of the Chengjiang Fossil Site project. This was the message which was sent by vice mayor Yang Yang from Smolny. We were at the Chengjiang Fossil Site where were located at 102°58'38" east longitude and 24°40'08" north latitude. There was a full four hours time difference between the two places. Ten minutes later, the Chengjiang Fossil Site was officially inscribed on the World Heritage list.

Meanwhile, over the square Chengjiang Fossil Site County Museum the masses people that participated in the celebration were singing and dancing, the major media reporters were busy broadcasting the live situation in the two places. On behalf of Yuxi, I gave the speech: "So many things came from the Chengjiang, and dreams become true. Dreams of yesterday have already become true, and tomorrows will become a reality at last!" Those sentences are from my creation "Cradle of Life" lyrics. Little fossil dream has an interesting and long story. It is imperative to tell the old story now, although I have mentioned earlier.

When France entered the period of the rise of capitalism, the government crazily expanded its powder and plundered resources outwards, it also dispatched a group of highly qualified natural scientists around the world for scientific study and military expeditions. In July 1904, J.Deprat (born in 1880) and H.Mansuy (born in 1857) who were ordered to take up his post as director of the Institute of geological survey of Indochina, made many trips to Chengjiang Yunnan China and Vietnam, Cambodia to conduct geological survey. 1907 – 1912, they co-authored and published a number of studies on Yunnan Geological Survey papers and books, the one that published in 1909 *Memorandum of eastern Yunnan Geology* was first announced to the world that the Chengjiang Maotianshan Mountain belongs to the Cambrian and they also found the Cambrian fossils. Unfortunately, this great discovery was craftily framed as scientific deception events by "fabricated fossil", and this event led to the outstanding geological scientists J.Deprat lose his position of the director of the Indochina Geological Survey. More unfortunately, they had no conditions to find Chengjiang animal fossils; all of their pursuit became illusory. Fortunately, Douglas Pall wrote an article "Burgess Shale a rival", highly praised the two outstanding geologists, and revealed the tragic hardships in geological history of science.

Early 1939, in order to escape the invading Japanese army's massacre, the thousands of teachers and students of Sun Yat-Sen University moved to Yunnan Chengjiang County. The Department of Geology was placed in Kuama Village. March 1940, famous geologist Professor He Chunsun and Bao Liyi came to the school to teach those students, and they also came to investigate Chengjiang Dongshan phosphate resources. 1942, Professor He Chunsun in *the Geological Report* No. 35published "Yunnan Chengjiang Dongshan phosphate Geology", which was first announced to the world Maotianshan's geological map and profile, and provided the basis of "There are fossils of lower organisms within the shale."and "German professor Mishi already have had the trilobite". As a distinguished honorary Professor, in his early years He Chunsun worked in the National Government Central Geological Survey, in 1943, with Bi Qingchang, he was ordered to organize the northwest branch of the Central Geological Survey, later in 1949 he came to National Taiwan University to teach geology. As so far, I don't have the evidence that He Chunsun, who was the founders of excavating Maotianshan mountain fossils, had read J.Deprat papers. However, Hou Xianguang, who firstly found the ammonite res fossils in Maotianshan, had already carefully studied J.Deprat's raw data, and followed instruction from Nanjing Maotianshan mountains. On July 1st, 1984, Hou completed their unfinished work, which has been hailed as one of the world's most amazing scientific discoveries of the twentieth century.

Since then, the Chengjiang Fossil Site became treasured place that attracts the attention from the home and abroad. 1987, Chengjiang County announced the "Interim Provisions on the Protection of the Chengjiang fossils of invertebrates", founded Chengjiang Fossil Site protection authorities. 1992, UNESCO inscribed the Chengjiang Fossil Site on the "Global Geological Sites List". 1997, the Yunnan provincial government designated 18 kilometers of land around the Chengjiang fossils as part of the nature reserve. July 1st,

2004, the former governor of Yunnan Province He Zhiqiang visited Maotianshan Mountain. As a mine engineer, he firstly put forward the concept of applying "the United Nations World Natural Heritage". The government of Yuxi city and Chengjiang country immediately organized preparatory work. September 5th, 2004, former Premier Wen Jiabao instructed at the Xinhua News Agency internal reference that: "to protect the Chengjiang Fossil Site, to protect the world's fossil treasure, to protect the natural heritage of great scientific value." Yunnan Governor Xu Rongkai instructed, "Resolutely implement Premier Wen's instructions, take effective measures to protect the Chengjiang fossils."

The government in Yuxi city and Chengjiang county held the Party committee meeting to deploy and implement Premier Wen and Governor Xu's instructions at once. Finally, in our place the government shut down many mines around the fossils, a total of 14 phosphate nine with 16.7 billion RMB value, invested nearly 100 million, return the grain plots to forestry 20,000 mu, deprecated more than one thousand of motorized boats that may pollute Fuxian Lake, and took the opportunity to pursue efforts to apply for World Heritage, the committee decided the mayor assistant Qing Weidong take the charge of it. November 2005, the People's Government of Yunnan Province presented Chengjiang Fossil Site Application for *World Heritage's referrals* to the State Council. In 2006, the Chinese government recommended "Chengjiang Fossil Site inscribed on the World Heritage Tentative List" to the United Nations. 2007, Vice Premier Zeng Peiyan and Yunnan Provincial party committee and government leaders visited Chengjiang fossils, and put forward higher requirements in order to protect the world's fossil treasure and "application for the World Heritage" work. April 2009, the city of Yuxi Municipal confirmed vice mayor Yang Yang to take charge of promoting Chengjiang Fossil Site application for the World Heritage. 2011, the Chinese government submitted application's matters to the United Nations, and the United Nations sent an expert team to assess the Chengjiang Fossil Site. On June 24th, 2012 (local time) 17:00, Vice Mayor Yang Yang led a group to St Petersburg Smolny to make the application for World Heritage's dream come true.

As I look back, scenes of the past are leaping before my eyes, as though they were something new. The odds are harsh, but also work miracles!

From July 1st, 2004 to July 1st, 2012, it took 8 years to spare no effort to protect Chengjiang Fossil Site. During this epic age, it created many new experiences. We should be inspired to work harder than before in the development of Chengjiang Fossil Site administrative departments to encounter the world. We are now shouldering our historical responsibility.

From July 1st, 1984 to July 1st, 2012, it took 28 years to build the enduring brand of this cradle of earth's life. During this epic age, prominent scientists such as Hou Xianguang, Chen Junyuan and Shu Degan have been fruitfully studied from Chengjiang fossils. We should be inspired to work harder than before in the development of research platform into the service of the scientists both from home and abroad. We welcome all talents to find the scientific value of Chengjiang Fossil Site.

From July, 1904 to July 1st, 2012, it took 108 years to explore the Chengjiang Fossil Site by the scientists such as J.Deprat, H.Mansuy, He Chunsun, etc. During this epic age, emotional and stirring stories came out of this preliminary period. We should be inspired to work harder than before in the development of Exhibition Hall, to convene the International Geological Conference. We would like to show the world the nature's greatest miracle, and to spread the scientific cosmic view.

The dream of fossils has been over hundred years. Glad tidings of the application for the list of world heritage were brought to everywhere. The story of the future may be like: recollections of the past, dreams for tomorrows. I have a dream to see the application for the list of world heritage could wake up the fossils, and the ancient livings could return to the big world.

后　记

写作完这本小册子，仿佛是上完了一次高等大学！我有缘结识了许多杰出的导师和优秀的同学，他们给予我大量的新知识和大力的帮助。在此，我对支持完成此书的全体师长和朋友，一并致以衷心的感谢！

Postscript

Finish writing this book, I seem to finish a college study! I've met many remarkable teachers and excellent students and they gave me abundant sources of new knowledge and great help. Sincere thanks to all the teachers and friends for supporting me with this book!